TRANSPARENT GOLD
透明的黄金

by
David Chen
南郭求败

Transparent Gold
透明的黄金

ISBN: 978-0-9861612-6-1

Claim the Earth series

《玩世》系列丛书

Fossil fuel used to be called “black gold”. Now we will show you how to extract another form of energy that we called “transparent gold” from the ambient air, and save the environment in the process.

化石燃料曾被人们称做 “黑色的黄金” 。本书将向您推荐一种新的提取能源的方法 - 开采 “透明的黄金” 。 “透明的黄金”的储量、开采范围、以及诱人的商业利益和前景与传统成语中说的“点石成金”差不多，也就是说：在我们周围的空气中到处都是。还有更令人惊喜的是，开采“透明的黄金”可以保护环境和维持生态平衡，同时造福子孙后代。

Fill a bottle full of water and then turn it upside down. How long does it take for that 1 liter of water (1 kg) to be pushed to an altitude of 1 meter? I didn’t have a 1-meter or larger bottle, but according to this picture, the deducted answer is close to instant (just about 0 second !!? for about 1/5 meter). The power and wattage of atmospheric pressure is stunning.

在一个可乐瓶中灌满水，然后在水中把它翻转过来。想一想 1 升的水被抛到空中 1 米高需要多少时间呢？从上图中的显示观察，推出的答案是：瞬间。哇！大气压代表的功率，马力，瓦特值竟然如此惊人、藏而不露。

We can now add a crazy balloon to conduct an experiment about those enclosed water in the coke bottle.

现在我们加一个不规则形状的气球，放入被封闭与大气圈隔离的水中做一个实验。

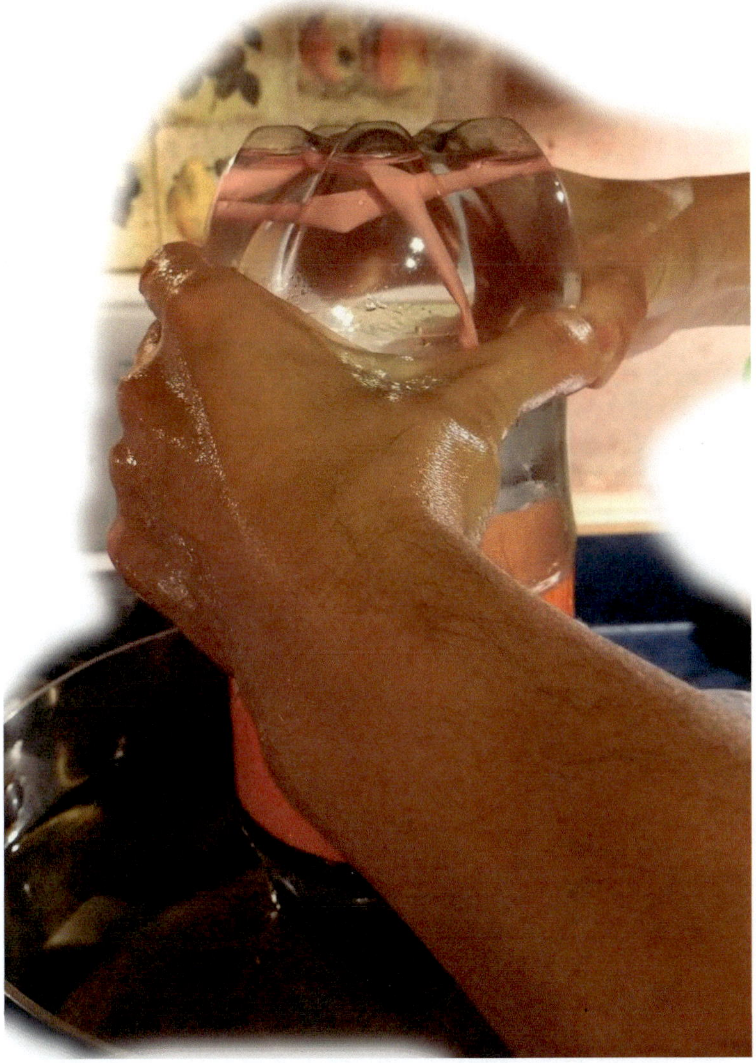

The balloon floated to the top of the bottle, signaling the laws of buoyance still apply in the enclosed water.

封闭体系内的水中,浮力仍然工作如常。

Within a confined segment of PVC pipe, if we fill the pipe and stand it up in a bucket full of water. . .

在一段一头封闭的聚氯乙烯管道中灌满水，然后将它立于一桶水中。。。

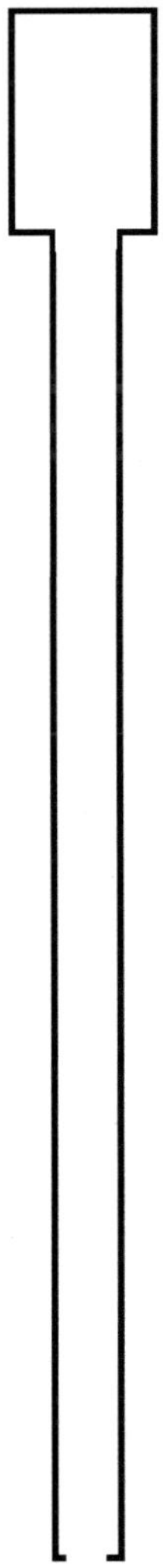

. . . almost instantly, the atmospheric pressure would have pushed the water to more than 1 meter high just like it did to those water in the coke bottle.

。。。像刚才在可乐瓶中一样，瞬间大气压就会把水在 PVC 管中推到 1 米的高度。

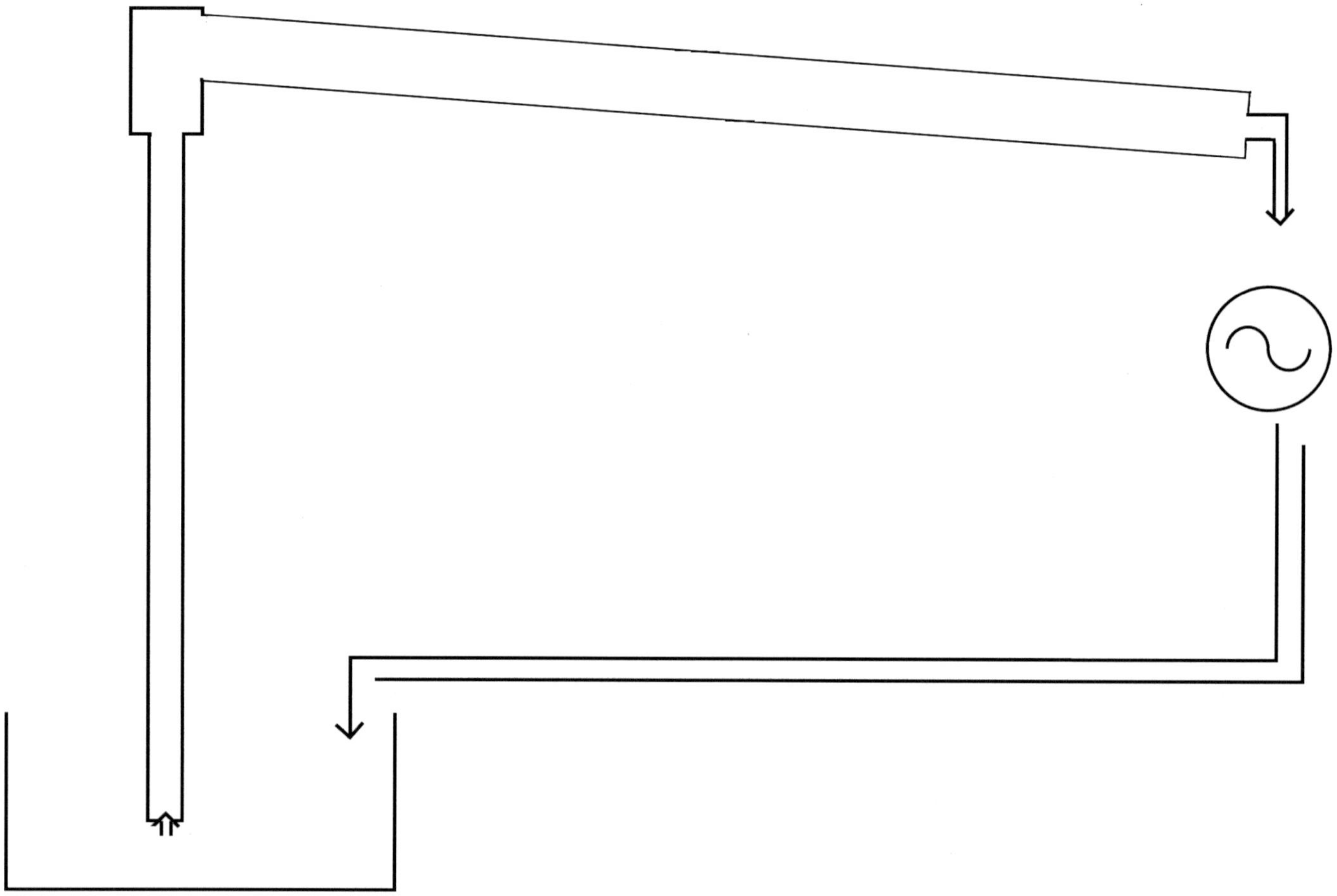

Now if we can figure out how to add an unbalanced siphoning drawing force to the right, and put a check valves to the left (optionally another one to the right end), we would get us a continuous flow of water that can generate electricity, for quite a long time.

接下来我们如果能造一个不平衡的虹吸管，不让水倒流，而是以惯性力将水“吸”往另一个出口流出的话，我们就建成了一个可以发电不止的瀑布了。

Here, it worth reemphasizing the century old discovery about energy and its omnipresence:

$e=mc^2$

wherever there's mass (fermion), there's energy

$e=h\nu$

wherever there's change, there's energy, even with emptiness (like photon, a type of massless boson particles)

The greatest 20th century advancements in science are: relativity and quantum physics, which led to the realization that all forms of forces are ultimately inertial forces (or fictitious forces) and all forms of energy are ultimately potential energies.

值得回顾一下，二十世纪人类科学意识的重大突破：我们周围能量无处不在。相对论和量子力学告诉我们说，1）有质量就有能量；2) 有变化就有能量。

所有的力，最终本质是惯性力。所有的能量，归根结底是势能。

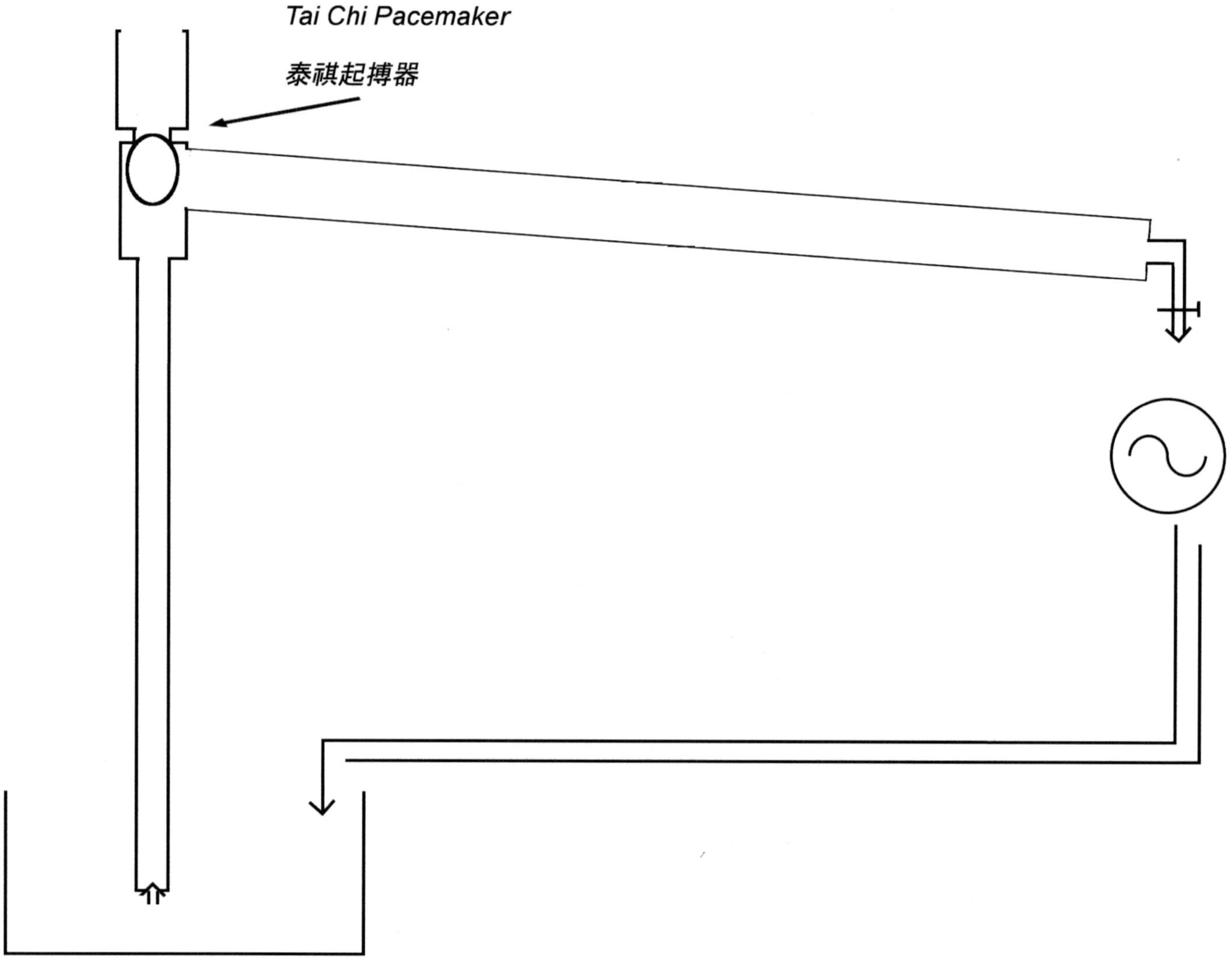

Even though the theory is great, practically there are still many difficulties to the above design when it comes to operation. How to fill such a system with water and turn it upside down to start the siphoning process? Also, since no machine can run perpetually, how to refill water to sustain the waterfall supported by such a circular siphoning process? The answer is to add extra containers and extra valves like in animal hearts. Oh, and don't forget to add an extra pacemaker device that is dubbed: "bubble valve" or "Tai Chi Pacemaker" as its official name.

理论上说得再好，也是光说不练假把式。实际操作起来，如何引发并持续大规模的虹吸现象是一个难题。让我们发明并隆重推出一种叫“泡泡阀门”或学名叫“泰祺起搏器”的神奇装置。

This is how we constructed our man-made waterfall and circular siphoning system. It only occupies a small corner of a basement.

这是我们造的人工瀑布的实图。空间上只占地下室的一个小角落。

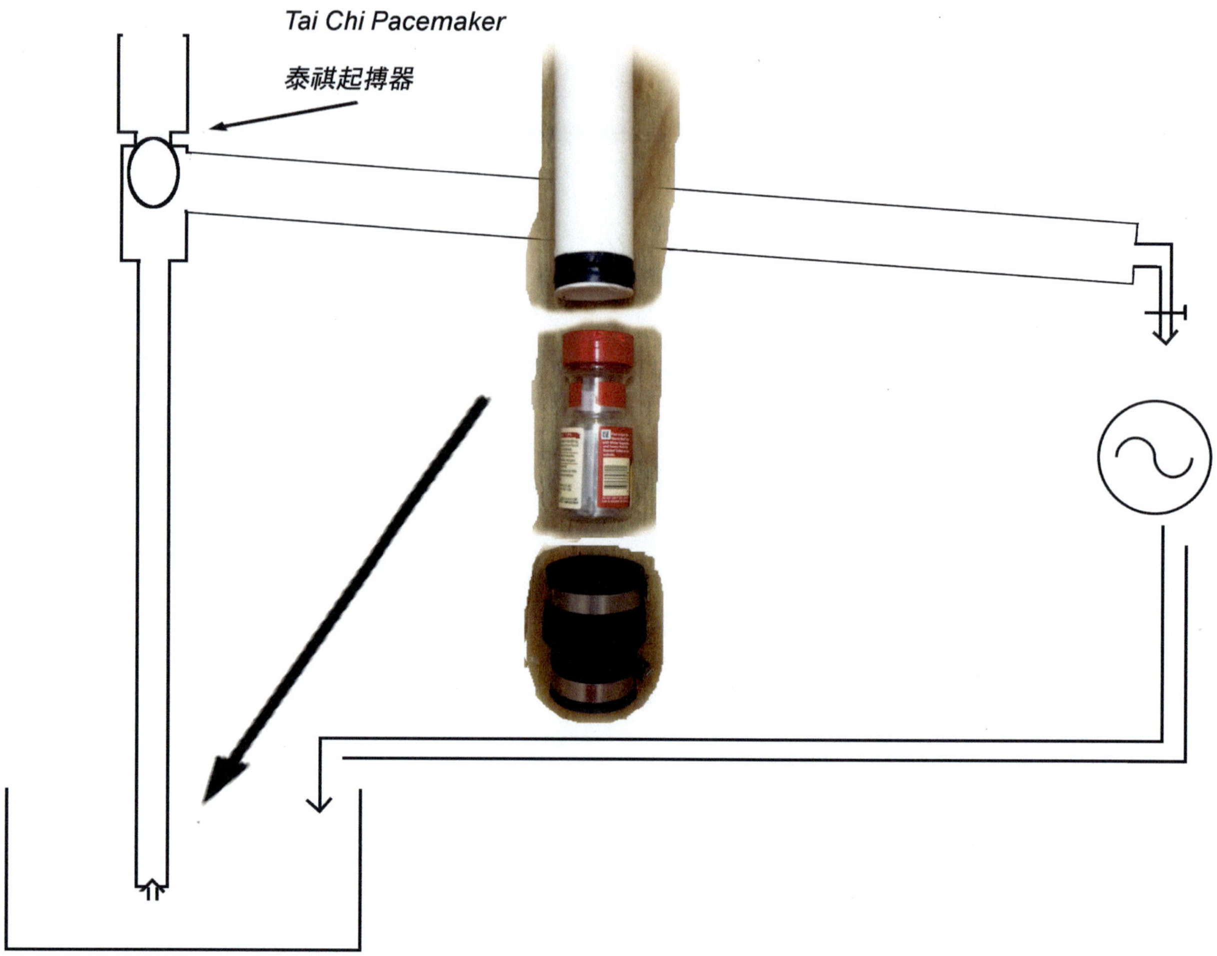

This is the check valve used in the project on the left side. It can be assembled with regular house hold items or recycled waste materials. We can fill the bottle with sand to adjust its pressure differences for the bottom check valve.

左下角的单向阀，可用密封的小瓶子装上沙来调节两个方向的压力差。在建造这个装置的过程中，许多废物、回收物都可以得到妥善的利用。

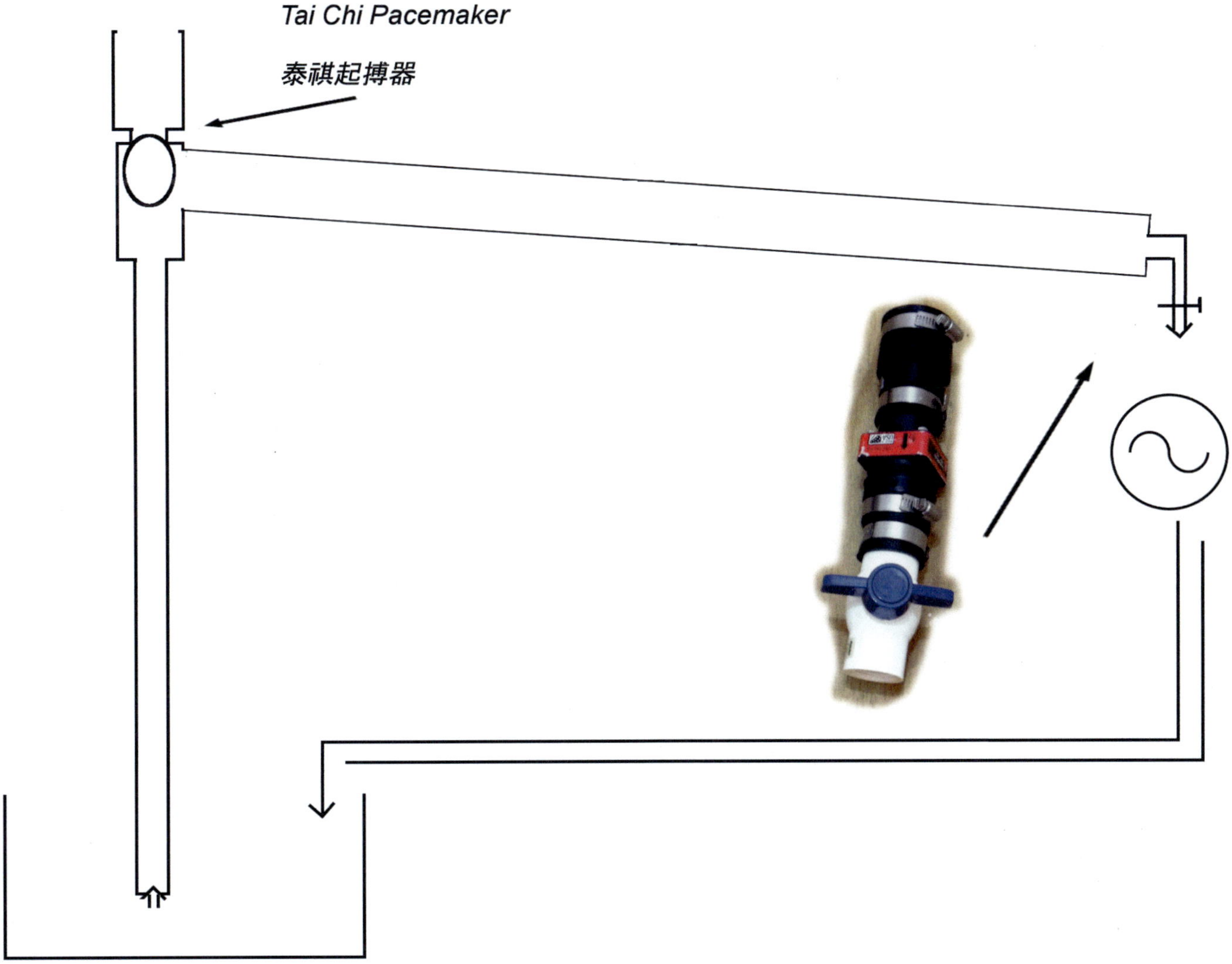

The right side of the system also has a check valve. For operational convenience, it is better to attach a faucet with it, to control the outflow of the waterfall.

装置的右边也有一个单向阀。为了操作方便还可以接一个水龙头，控制出水流量。

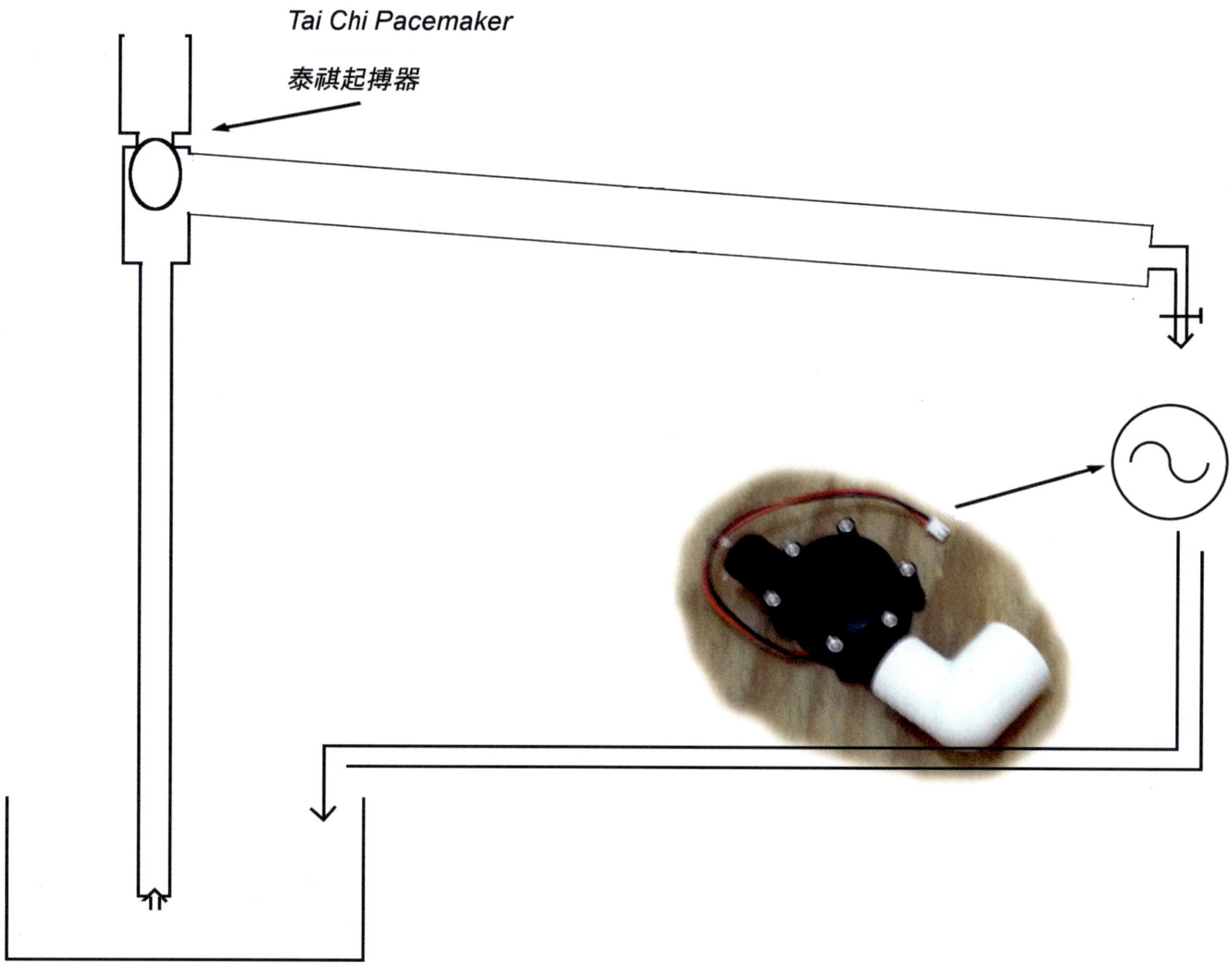

The generator can be small or large, depending on the scale of a waterfall that would be built to power it.

微型水力发电机可大可小，视瀑布的势能储量大小而定。

Purchased or home-made materials can be used to connect the drainage areas in and out of the generator.

发电机的进水管和出水管可利用商用或家用的材料来连接。

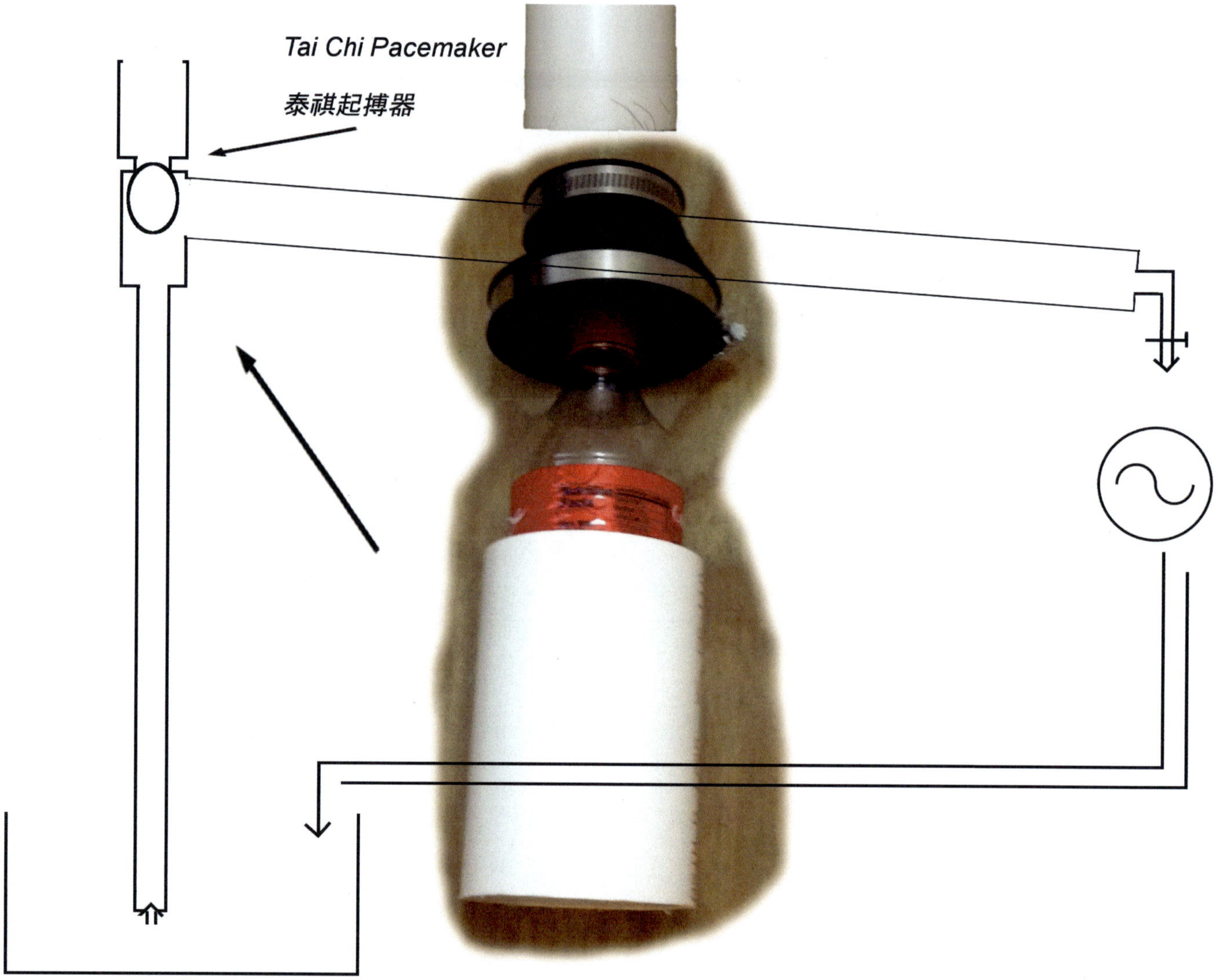

The all-important pacemaker part does not have to be a round rubber ball as depicted in the figure. It could be as simple as a properly-sized coke bottle.

最重要的起搏器部件不一定要用马桶橡皮球。更常见、易得的材料是一个小可乐瓶。

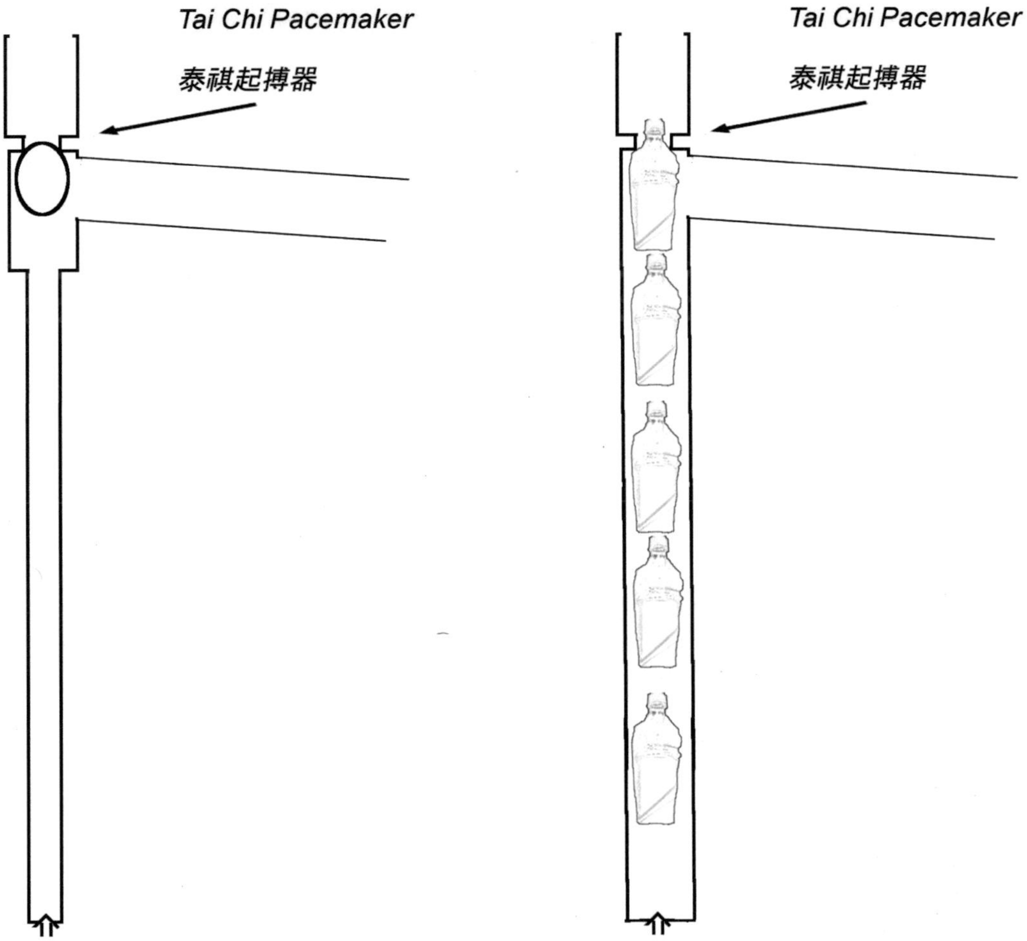

It could also be a series of coke bottles with plenty of air filled and sealed in.

用多个串联密封的可乐瓶也可以工作，而且有可能提高功率。

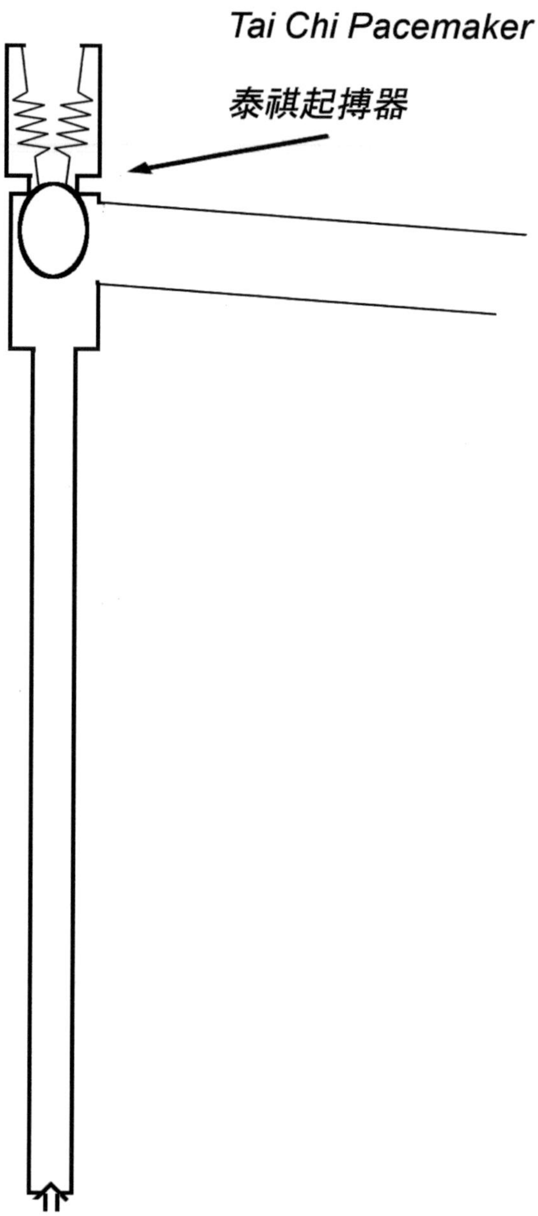

For the Tai Chi pacemaker, this is another alternative design that is still going to work. The design of this part should take into considerations of: 1) producing enough power and 2) maintaining operational longevity of the system.

关于泰祺起搏器，还有许多其他的设计方案。考虑的关键因素是要平衡装置的以下两种需要：1）增加功率 2）延长运营寿命

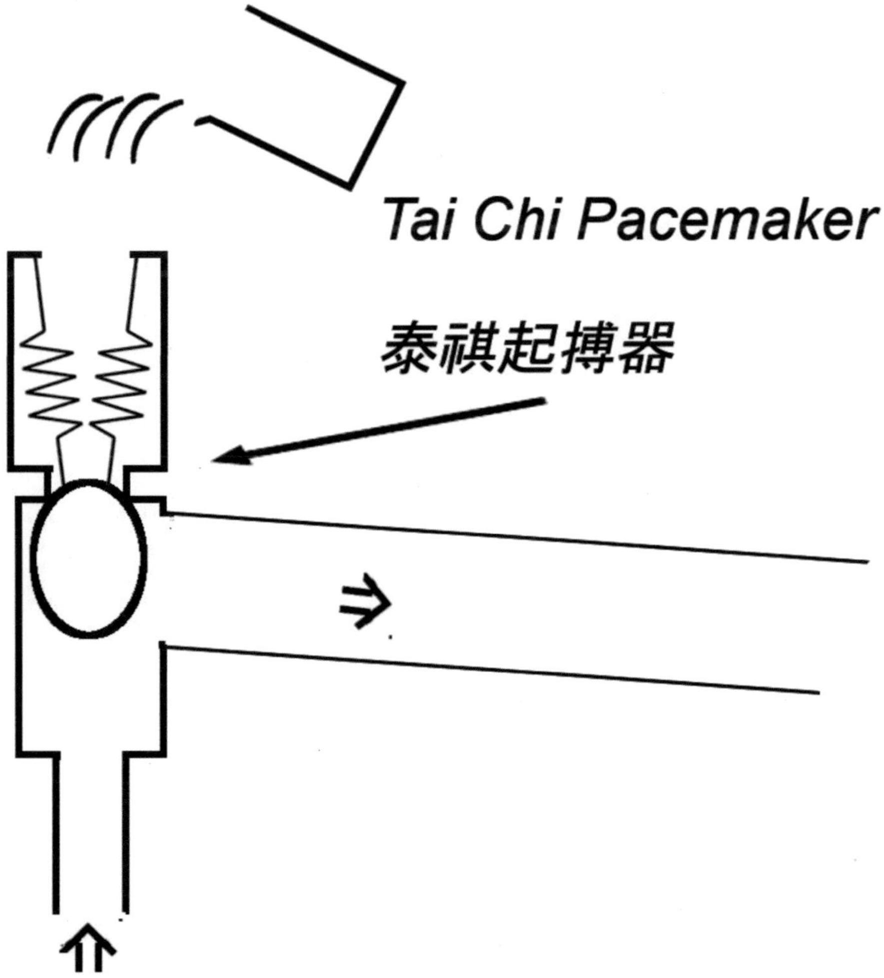

With this crucial part of the system built, just pour in water through the pacemaker and the indoor/household chain-reaction waterfall will eventually start. Every home would be able to have a hydraulic power station without building a dam. Every household should be able to enjoy plenty of renewable energy regardless of outside wind speed or sunshine intensity.

关键部位竣工了以后，只要加水，链式反应的瀑布就可以启动了。不需筑水坝，每家都可以建水电站。风力、和日照强度不够的家庭也可以享受廉价清洁的可再生能源。

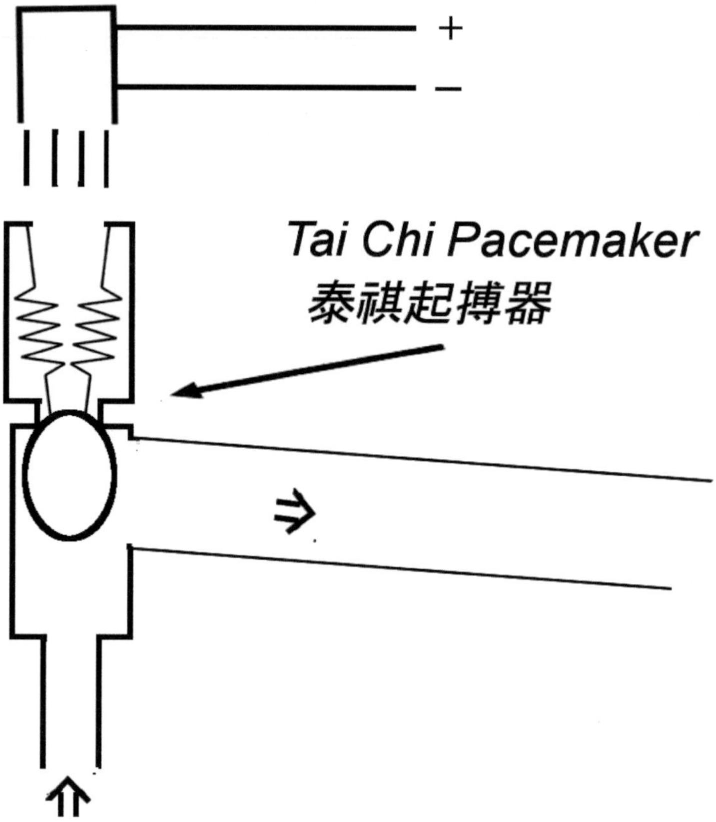

While it is a rule of the natural laws (The 1st Law of Thermal Dynamics) that perpetual machine cannot be built, it is not bad at all to build a machine that can run for 48, 24, or even 12, 6 hours after pouring into enough water every now and then to start or restart . . .

Better yet, the mechanism of pouring in water may be automated. To install a dehumidifier above the pacemaker and then connect the dehumidifier to the power source should do the trick.

热力学第一定律说永动机是不可能造出来的。但是加一次水就能转 48，24 甚至 12 或 6 小时的机器，我们觉得还是很不错的。

如果从空气中吸湿、加水的过程也能自动化了以后，我们就觉得更不错了。

One can buy a commercial dehumidifier that runs on AC. It is also possible and quite simple to build a crude dehumidifier out of any electrical fans or even discarded computer heat dissipation fans.

With the same mechanism as in refrigerators, condensed air when evaporated, can generally take heat away from surrounding environment and lower the temperature in a small region, just as described in the thought experiment known as “Maxwell’s daemon”.

With cool air blowing pass the surface of a piece of metal, condensation water drops will drip down, because there’s always small amount of moisture in the atmosphere in room temperature.

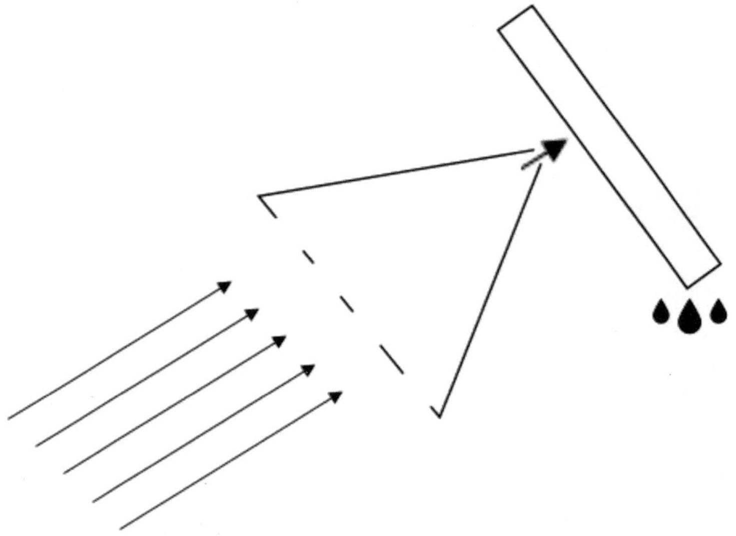

吸湿器可以到商店买一个现成的用交流电的。用一般的电扇、冷却扇自制一个简单的也很容易。

从电冰箱的工作原理我们知道，压缩空气挥发时会带走热量，降低周围温度。这种现象被总结为“麦克斯韦尔的妖”的原理。

当冷空气吹到金属表面时，室温空气中的水分就会凝结、滴落了。

We use these items to build our dehumidifier based on the idea of condensation.

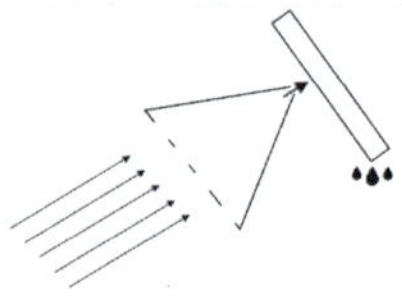

可用这些废旧物品安装一台简陋的的吸湿器。

A side effect of generating electricity from atmosphere is that water and precipitation can be generated from moistures in the air around us. Such a process is like reproducing a complete large-scaled water circulation process in nature (in atmosphere), within a reduced small realm of one’s basement. This byproduct of power generation can also help solve the problem of drinking water shortage in many parts of the world.

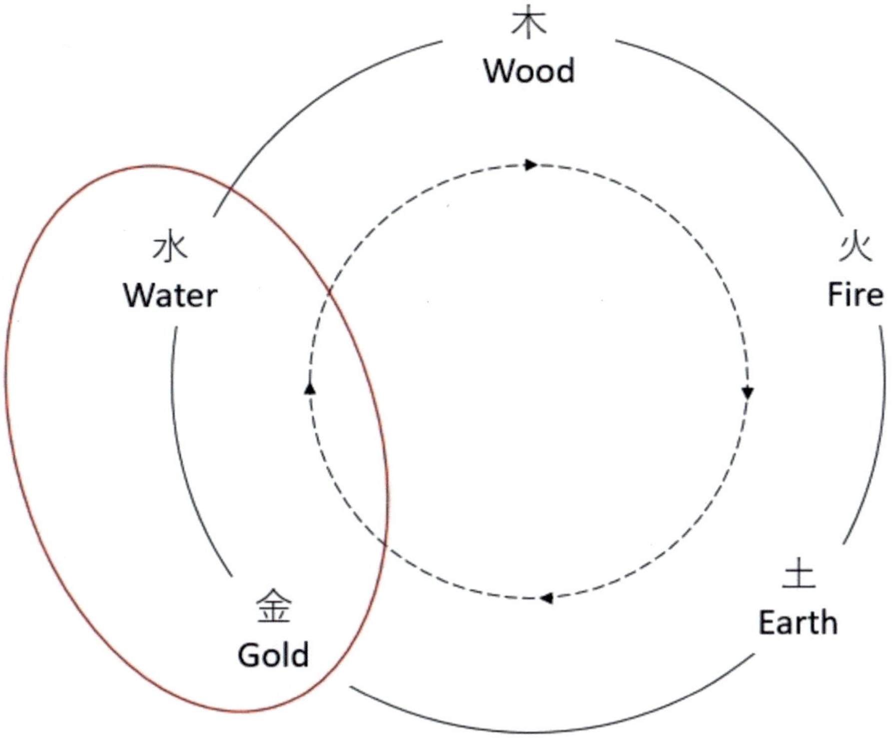

“木生火 火生土 土生金 金生水 水生木”，用大气圈中的水循环，在微观的小范围内重演以发电，得到的另一个意外收获是：可以给世上许多缺水的地区带来饮用水。

Transparent gold is the type of money that can solve energy, environmental and water shortage problem in one form of currency. As "gold", it has traits and characteristics of commodity currency, hard currency, fiat money and international exchange reserve currencies. It can also bring essence of life - i.e. energy - to areas that were hitherto not possible.

透明的黄金是一种在能源、环境、水资源等多行业的通用货币。由于固有的含"金"量，它具有商品货币、贵金属货币、法定货币，以及国际储备货币所具有的性质。更重要的是它能给以前荒芜不毛、无法开垦利用的地区带来生命的本质 — 能量 — 原来祂居然是无处不在的！？。

Made in the USA
Monee, IL
16 November 2026

68663815R00017